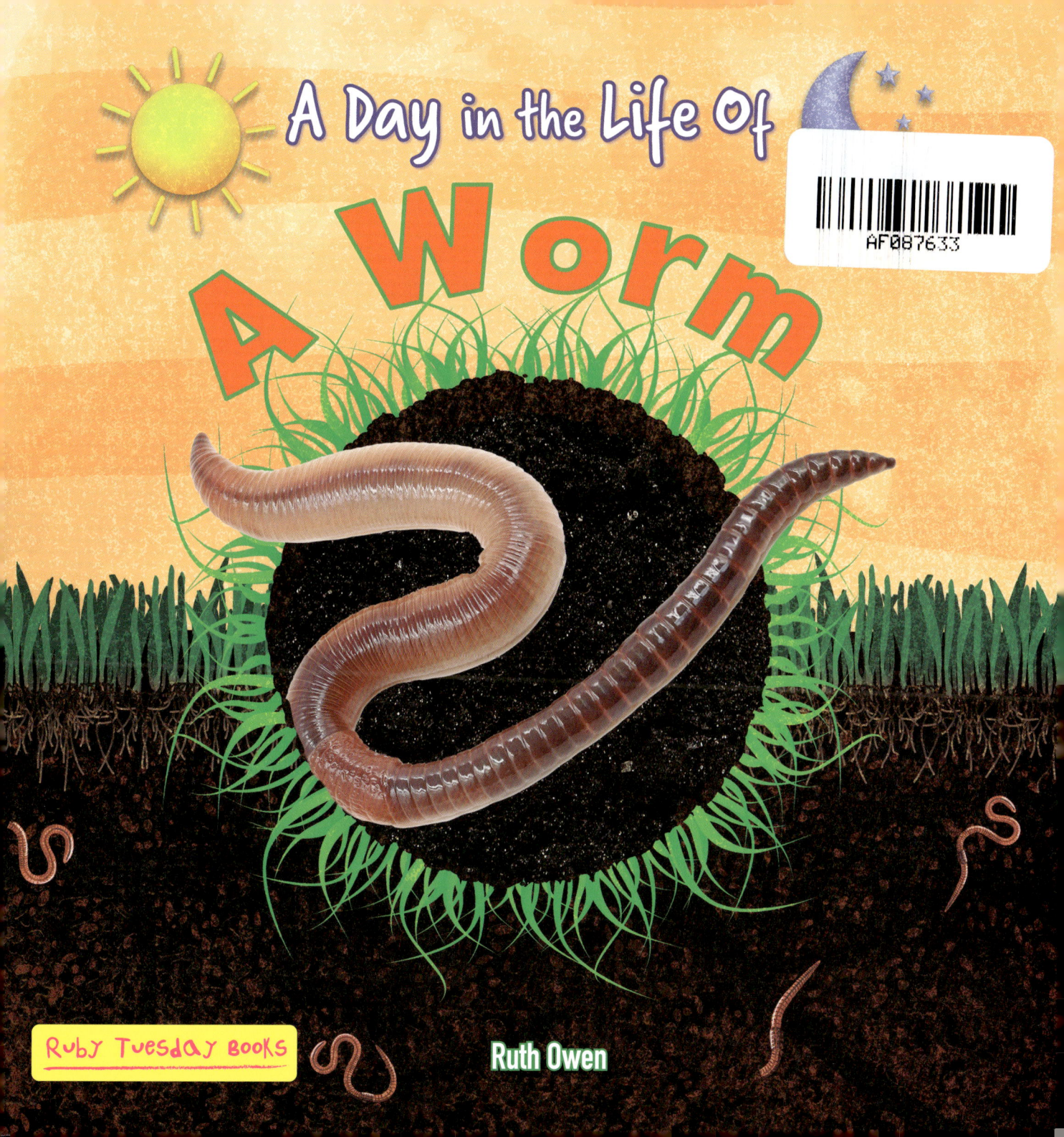

Published in 2025 by Ruby Tuesday Books Ltd.

Copyright © 2025 Ruby Tuesday Books Ltd.

All rights reserved. No part of this publication may be reproduced in whole or in part, stored in any retrieval system, or transmitted in any form or by any means, electronic, mechanical, photocopying, recording, or otherwise, without written permission from the publisher.

Editor: Mark J. Sachner
Design: Tammy West
Production: John Lingham

Photo Credits:
Alamy: 9 (Auscape International), 14 (pq pictures); Dreamstime: 7 (Razvan Cornel Constantin); Dwight Kuhn: 11, 12T, 19; FLPA: 15; Nature Picture Library: 10 (Laurent Geslin), 17 (Solvin Zankl); Shutterstock: Cover (MakroBetz), 4T (Alexander Raths), 5 (D. Kucharski K. Kucharska), 6 (golf bress), 8 (Eviart), 12B (AliaksaB), 13 (neenawat khenyothaa), 18T (Tomasz Klejdysz), 18B (schankz), 20T (Peter Kniez), 20B (Wirestock Creators), 21T (Evan Lorne), 21B (Serhii Ivashchuk), 22 (GTW, kungfu01, & Kaca Skokanova), 23 (Majna, Ayman Haykal, Tama2u, & krolya25), 24 (Tob1900).

ISBN 978-1-78856-437-3

Printed in Poland by L&C Printing Group

www.rubytuesdaybooks.com

CONTENTS

It's Cool Underground! 4

Glossary 22

Index 24

It's Cool Underground!

It's a hot, sunny day in a garden.

But in an earthworm's underground world, it's always dark, damp and cool.

If the worm's skin gets dry, it will die!

So it stays out of the hot sun.

The worm's body is soft, but it can push through solid soil.

It has hair-like **bristles** on its skin.

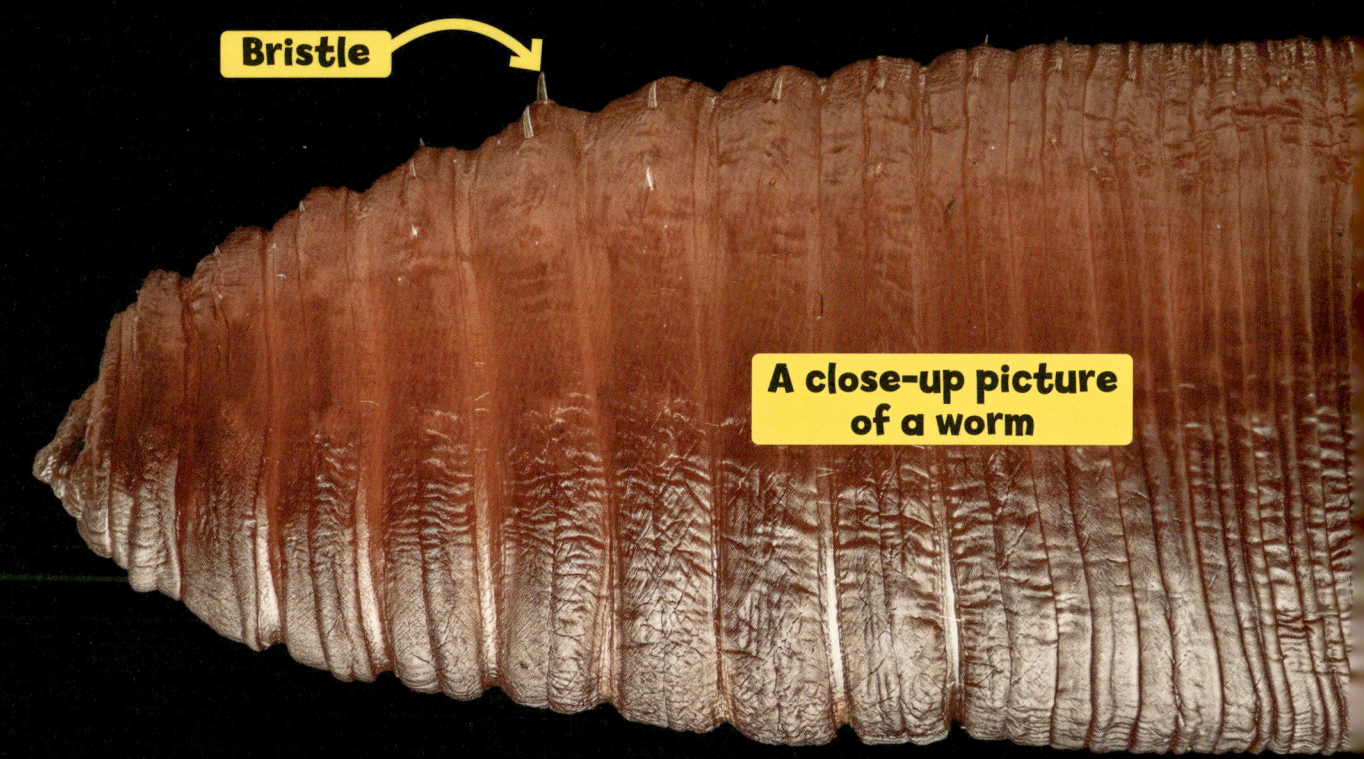

Bristle

A close-up picture of a worm

The bristles help the worm push forward.

The worm is very helpful to the garden plants. How?

Worms make little holes and tunnels in the soil.

Rainwater trickles through the holes and tunnels.

This helps water get to the roots of plants.

When evening comes, the worm wriggles from its underground world.

It's time to find food!

The worm eats **rotting** leaves.

Dead, rotting leaf

Worms help clean up dead plants.

The worm eats rotting fruit and vegetables.

A rotting pumpkin

The worm makes poo called castings.

Worm castings

Castings are full of good stuff.

The castings become part of the soil, and help feed plants.

When morning comes, the worm wriggles back into the soil.

Worm's tail

But a hungry bird grabs the worm's tail!

The bird pulls

and pulls

and pulls.

The worm's bristles hold on tight to the soil.

BOING!

The bird lets go, and the worm gets away.

But there are hungry **predators** underground, too.

A mole is hunting for worms in the soil.

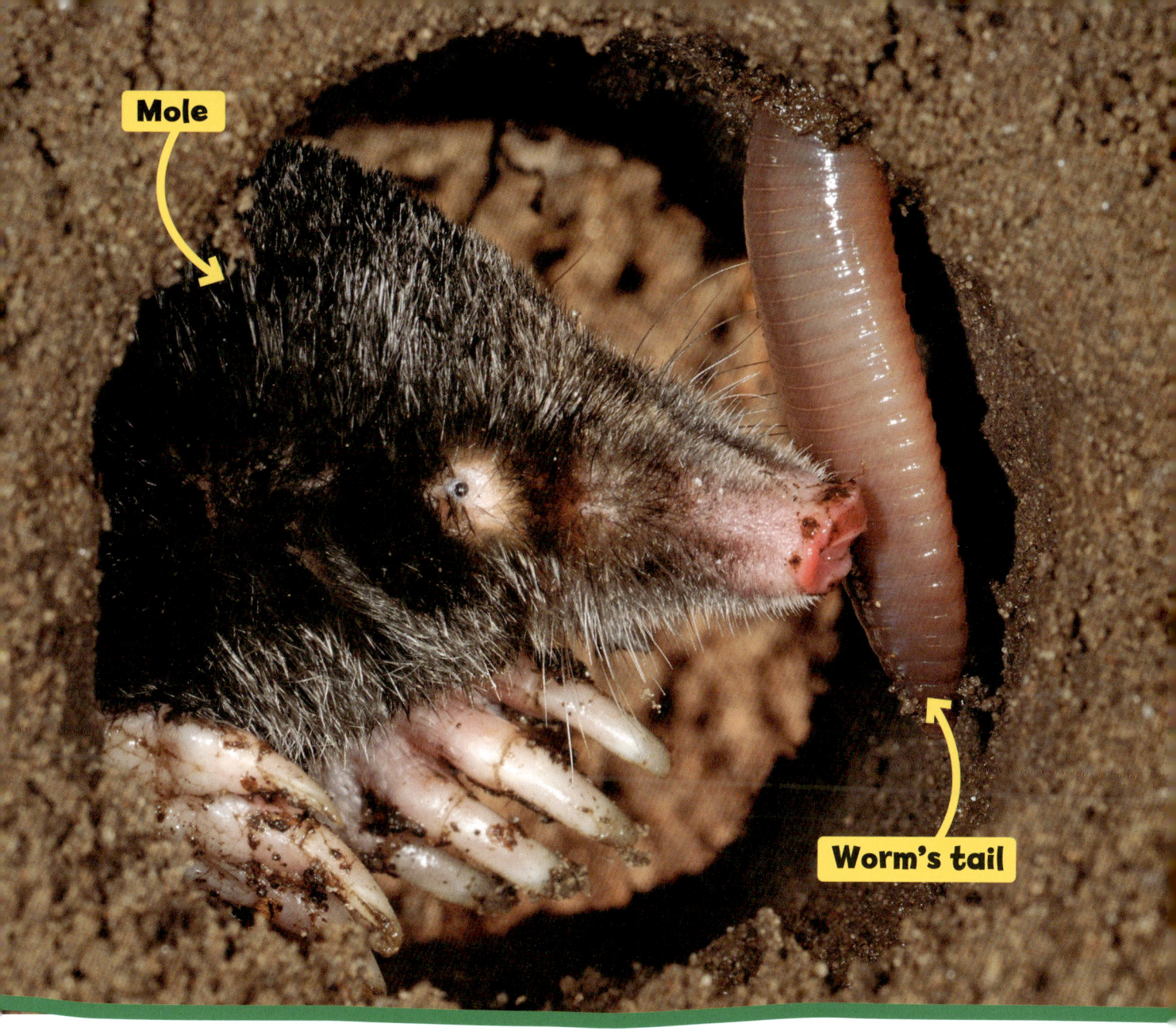

Just in time, the worm gets away!

Today, the worm makes a **cocoon**.

There are eggs inside.

Cocoon

The cocoon comes from here.

A baby worm hatches from each egg.

It wriggles from the cocoon.

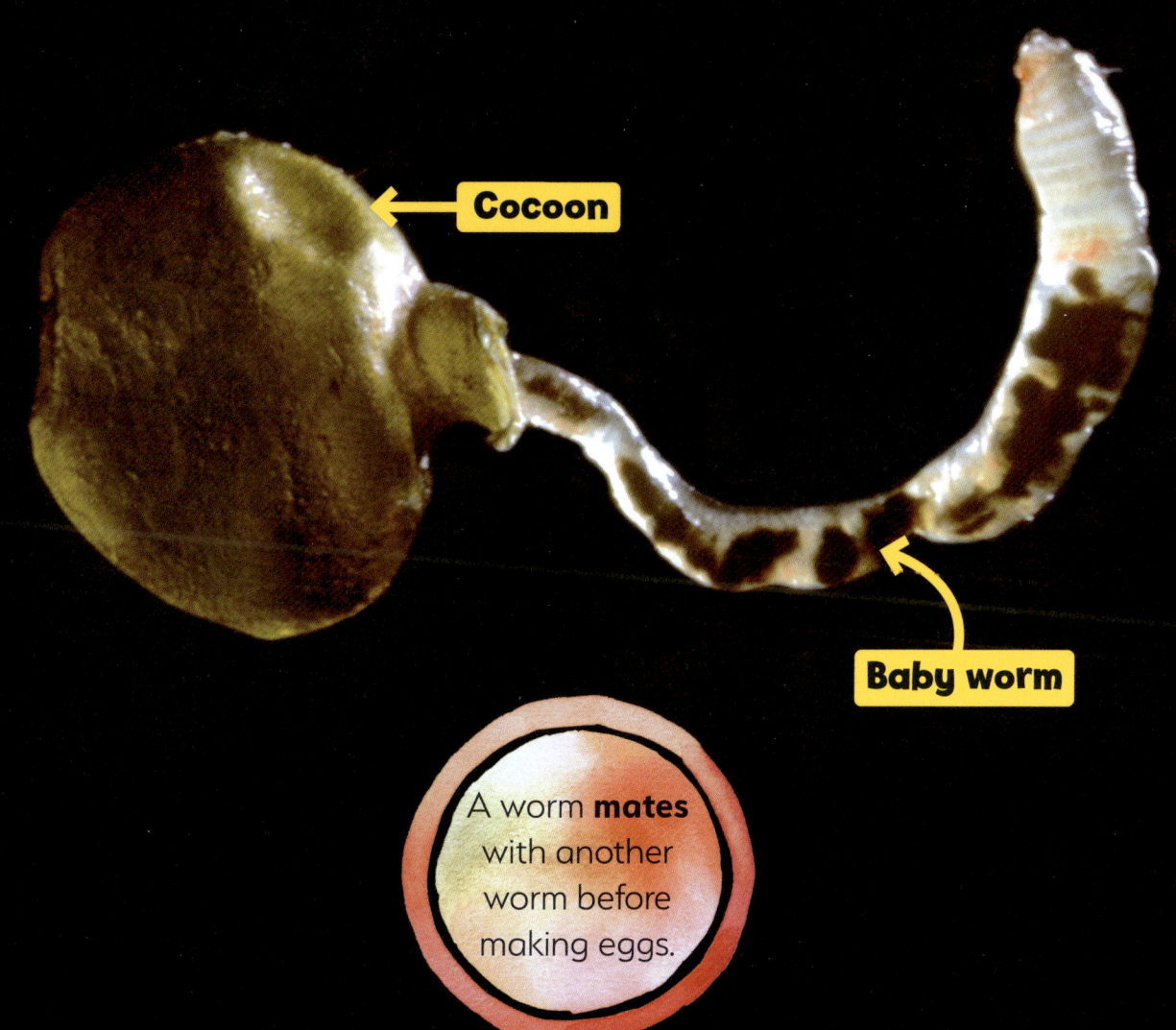

Cocoon

Baby worm

A worm **mates** with another worm before making eggs.

Suddenly, the worm's underground world is moving and shaking!

Up Up Up

Someone digs up the worm.

They put the worm on a **compost heap**.

Compost heap

Worms eat rotting stuff on a compost heap.

It's the perfect new home for a worm!

Glossary

bristle
A short, stiff hair on an animal or person's body.

Caterpillar cocoon

cocoon
A covering that protects eggs or a young insect, such as a caterpillar.

compost heap
A box or heap of dead plants, old vegetables, eggshells and other matter that rots and becomes new soil.

mate
To come together to produce young.

predator
An animal that hunts and eats other animals.

rotting
Breaking down and becoming mouldy. Rotting leaves and fruits become part of the soil.

Index

B
bristles 7, 15

C
castings 13
cocoons 18–19

E
earthworm's body 5, 6–7, 14–15, 18
eggs 18–19

F
food and eating 10–11, 12, 21

M
mating 19
movement 6–7, 8, 10, 14

P
predators 15, 16–17

T
tunnels 8–9